Human Cloning

Written By
Acie Cargill

Copyright © 2019 by Acie Cargill.

All rights reserved.

Published by Acie Cargill

aciecargill@gmail.com

http://aciecargill.com

ISBN: 9781699323779

Imprint: Independently published

Formatted - Brenda Van Niekerk

brenda@triomarketers.com

Website Design - Brenda Van Niekerk

http://triomarketers.com

Synopsis

This 7000 word booklet begins with a brief introduction to the field and theory of human genetics. The DNA inside a cell in the form of chromosomes that contain all the genes that make us into who we are. Then how it works with the sperm and the egg in normal reproduction but with cloning how the embryo is prepared. It will be then carried by a surrogate in her uterus until the baby clone is born.

The book describes three types of clones that will probably be in common use someday. Comatose clones have most of their brains inactivated in the embryo so it will just lay in a comatose state until it needs to be sacrificed for spare parts. Another type of clone is the robotic clone which has most of the brain and can function somewhat but without the higher portions of the human ego. The third type of clone is the intact brain clone. It is a complete person and will not be killed for parts or used in any way. It will have its own free will and make all its own choices for its life.

This book is a basic description of the field of cloning. It is complex in a way, but the material is presented plainly and clearly and fairly understandable for the non-scientist to understand. There is some mention of the morality, the legality, and the sexuality of cloning. This is a basic introduction to the theory of cloning.

About the Author

Acie Cargill is a poet, a songwriter, and a prose writer. He studied poetry with USA Poet Laureate Mark Strand and Illinois Poet Laureate Gwendolyn Brooks. He studied novel writing with Thomas Berger, who wrote Little Big Man (that Arthur Penn made into a movie with Dustin Hoffman in the lead role). Cargill also studied journalism with instructor Jean Daily. His work is a synthesis of all these styles.

He is a member of American Mensa and formerly Edited the Mensa Journal of Poetry. He also is a member of the Grammy Association, and The US Quill and Scroll Society.

Cargill is a vegetarian, a former holistic physician, a musical performer on a variety of instruments, an environmental activist, a lecturer, medical reviewer, a lover, and a seer.

Website

http://aciecargill.com

Contact

aciecargill@gmail.com

Other Books Written by the Author:

Puerto Rico

Aberrations

Chronicles

Terrorism

Modern Love

Ends and odds

Illiana: The Border Area Between Illinois and Indiana

Pullman

Che and Fidel - A Reading Play of the Cuban Revolution

Celia Sanchez - A Play of the Cuban Revolution

Paschke - A Play

Gwendolyn Brooks: A Play

Rasputin - A Play

Nietzsche - A Play

Bob Dylan, The Early Years - A Musical Play

Michael Jackson - A Play

Einstein - A Biographical Play

El Chapo - A Play In 3 Acts

Raisins and Roaches - A Three Month Diary of a Crack Addict

Susan B. Anthony - A Biographical Play

Kankakee

Harriet Tubman - A Biographical Play

Tesla - A Biographical Play

Vegan Saint - A Play in 3 Acts

Martin Luther King, Jr - A Play

Great Migration: A Play in 3 Acts

George Pullman - A Play in Three Acts

Frederick Douglass - A Biographical Play

Freud - A Biographical Play in 3 Acts

The Underground Railroad - An Educational Play

Payton, Jordan, Ali - A Biographical Play

Mr. Nobody - A Play

The Kid From Left Field - A Play

Puerto Rico, A Dream of Independence - A Play in 3 Acts

Crack Madness - A Monologue Play

Johnny Appleseed - A Family Play

Dr. Jekll and Mr. Hyde - A Modernized Play

Obama - Obama - A Play In 3 Acts

Will Rogers - A Biographical Monologue

Merle Haggard - A Biographical Monologue

Mother Teresa - A Biographical Monologue

Gwendolyn Brooks - A Biographical Monologue

Love Life of Susan B. Anthony - A Monologue Play

Sojourner Truth - A Biographical Monologue plus Narrator

Harriet Tubman and The Underground Railroad - A Play

Helen Keller, Words and Wisdom - A Biographical Play

Eugene Debs and the 1894 Pullman Strike - A Play

The Rising - A Play

Walt Disney - A Biographical One Act Play

The Experiments of Dr. Victor Frankenstein - A Play - Based on the novel by Mary Shelley

Karl Marx - A One Act Play

Martin Luther at The Diet of Worms - A One Act Play

Martin Luther King: Monologue and Narrator Play

Frederick Douglass - Monologue and Narrator Play

Kaepernick - A One Act Play

Settling South Holland - A Play In 2 Acts

Kaepernick - A Full-Length Play

My Son Died From An Overdose - A Play

Overdose - A One Act Play

Always a Marine First

Erotic Muslim Polygamy

George Dolton's Bridge to Freedom Underground Railroad - A One-Act Play

Greta Thunberg - A One-Act Play About Climate Change

A Brief History of the Philippines

Goat With No Horns - Voodoo Cannibals in Haiti

Johnny Cash - Monologue Play

Muhammad Words Of Wisdom

Jesus Words Of Wisdom

Bob Hope - Biographical Monologue

The Cargills of Graves County, Ky

Keith Raniere and the NXIVM Sex Club

Words of Wisdom – Native Americans, Ancient Greeks, Buddha and African-Americans

Words of Wisdom – Mark Twain, Benjamin Franklin, Shakespeare and Solomon

The Trial of Eddie Gallagher, Navy SEAL

Climate Crisis - A Plan to Prevent Future Flooding

Yukio Mishima - Life, Death, Hara Kiri

My "Cuzin Willie" Nelson - A Biographical Monologue

The World's Most Amazing Person, Elon Musk

The Beatles: Early Years - A One-Act Play

Red Summer Race Riot Chicago 1919 - Eyewitnesses John Harris and Ida B. Wells

Jeffrey Epstein - Illicit Kicks and Retribution

Gandhi - A Brief Biography

Jeffrey Epstein - Death Controversy

Greta Thunberg - Coming to America

Jeffrey Epstein Honeypots - Wealth, Women, and Girls

Tom Dreesen - Monologue Play

Roswell 1947

Greta Thunberg - How Dare You!

Suicide - Teen Suicide Prevention

Table of Contents

1. Introduction 1
2. Human cloning 4
3. Growing the clone and harvesting organs 9
4. An unusual request 12
5. Cloning for incest or pedophilia 15
6. Right or wrong? 18
7. Cloning multiples 20
8. Clone fantasies 23
9. Robotic clones 26
10. Clone reproduction 29

1. Introduction

Here is a very brief simplified introduction to genetics to help the reader understand what this book is all about. Our bodies inside and outside are made of cells. Trillions of cells. Skin cells, muscle cells, nerve cells, fat cells. Many types of cells and inside each cell is a nucleus that contains chromosomes which are long strands of DNA. DNA is a very complex chemical that is passed from your parents and as the cells divide, the DNA divides and goes into every cell in the body. Trillions of cells. Yep, we are very complicated creations.

Our chromosomes are half from our mothers and half from our fathers. An egg and a sperm each have half the number of chromosomes needed in each cell and when they join the egg is considered fertilized and will begin to divide and start to form a person. As the fertilized egg divides and multiplies, the chromosome also multiply, and go into every cell of the developing baby. All cells have the same chromosomes but not all are activated and that is what will make the various types of cells different from each other. For instance, muscle cells or skin cells. They have the same DNA chromosomes, but different parts of the chromosomes are activated and that is why skin is different from muscle.

Certain segments of the chromosomes are called genes which make us who we are physically. It is how we are different from everyone else. We are all unique because we have our own genes. We all have the same number of chromosomes made of DNA, but there are an infinite variety of combinations of the genes, so no two people are exactly the same.

A clone is made by taking the chromosomes from someone's cell. Any cell, and putting the chromosomes into a human egg called an ovum, But the ovum has been prepared by taking out the egg's own chromosomes so that all the chromosomes in the egg will now be from the cell donor and it will divide and form a baby. A person. But all the chromosomes will be from one person. Not half from a mother and half from a father which is normal reproduction. A clone will be an exact copy of the person who donated the cell that provided those chromosomes for the clone. All the chromosomes in a clone came from one person, not a mixture of a mother and father.

DNA is a very complex combination of chemicals that are used to preserve the identity of a cell as it divides. From one generation to another the DNA stays the same. It is how we inherit who we are. Same with all living organisms. Therefore, the DNA preserves our identity. Scientists study the DNA of all living organisms and especially humans and a lot is known as to what segment on the chromosomes have what function. What segments control the development of different structures. It is a very complex process. I am trying to keep this explanation basic and hopefully you can understand what kind of work that I do.

I am a molecular scientist and I work with high precision procedures to understand the chromosomes and what different genes do. It is not easy, but it can be done. 65 years ago we didn't even know what the function of DNA even was. We knew it was in the cells, but we didn't understand how it worked. We still don't know everything about DNA, but we know a lot, and more is discovered every day.

2. Human cloning

There are many aspects to cloning humans. That means making an exact duplicate of a person. And yes, cloning is very possible and in my opinion, it has already been done and clones are at the present time in some secret laboratories growing older. Maturing. After all they start as a tiny embryo and go through all the growth stages of a baby in normal reproduction and then grow slowly just like a normal person. As far as I know, at present there is no way to make the clone grow faster even though that would be very advantageous in many situations.

Why would someone want a clone created? It would be quite expensive for a normal person, but really not very expensive for a very wealthy person. Wealth is relative and prices are relative. What seems like a high price to one person would be a drop in the bucket for a very wealthy person. And there are a lot of wealthy people in the world. More every year from wealth growth and inheritances.

A clone grown for spare parts might add many years to a person's life. A new heart. A new liver. Most any organs and body structures can be replaced if needed. That is the reason for my own personal involvement in human cloning. Spare parts.

My price is one million dollars. If it sounds like a lot of money to you then my services are not for you. That price is cheap for what you get and that is not the only cost. It is just my fee is one million dollars. I will create an embryo of a clone that matches you perfectly. You have to provide a surrogate who will carry the developing child and maybe also be the mother to it after it is born. It is like any other baby. It needs to be cared for. The time necessary for caring for the child gets progressively less as time goes on. Just like with a normal person.

Actually, it would be cost efficient to do two or three clones at once. Keep them in the same laboratory facility and one person can care for 2 or 3 clones just as easy. My fee is one million dollars for each clone. My work is with the DNA and creating the embryo. It is the same work for each. Actually, it is not a lot of work. What you would be paying for is my training and high technical skills. And also a certain amount extra because what I will be doing is highly illegal.

Why is it illegal? Are we hurting anything? No. Are we stealing? No. I am creating from scratch. You and I would be the only ones involved. You are the donor of some DNA for me to work with. Of course, we will have to purchase some eggs from an egg donor. She has millions and won't miss the few that we remove. She will be well paid and for a couple of months, a physician will have to administer some of the proper hormones to her to increase her production of eggs and one day gather them from her secretions. Then she can walkout with a nice amount of money for her efforts.

It doesn't matter what she looks like. The clone will not have very much of her DNA. Just some specialized genes in her mitochondria. They are tiny energy producing structures in every cell, including ova, which are her eggs. And yes, these mitochondria do contain a small amount of DNA genes that will be passed to the clone.

It is not known if those mitochondrial genes will affect the clone's organs to be used as spare parts. My guess, is that those genes are too small to matter, but no one knows for sure because this has never been done with humans. At least not publicly. Anyway, it would be almost impossible to remove every mitochondria from the egg. They are so small. So yes, there will be some DNA in the clone from the egg donor. Not much. Probably not enough to cause an immune reaction if an organ is transplanted into the owner of the clone. Yes, owner. The clone will be owned by the person paying for my services and the care of the clone.

There is more than one type of clone. The type I work with, or I should say I am willing to work with are called comatose clones. I have disabled the genes necessary for the brain to develop so these clones have no sensations and no thought. They are kept alive by their brain stems like many comatose patients. The reason they have no brain is because they are growing to become spare parts for the person who donated the original cell from how we made the clone. And yes, this type of clone does have an owner. They are helpless and totally dependent on the owner to keep them alive, probably in a laboratory setting.

The other general type of clone, I call the brain intact clone. These clones grow to be a thinking person. They should not be killed for spare parts because they are complete people with their own lives to live. They have no owner. They have a person responsible that would be more like a parent until the clone is old enough to live on its own.

To make the clone I will need the chromosomes from a cell donor and an egg from an egg donor. I am going to take out the nuclei in her eggs that contains most of her DNA, because we don't need her DNA. I will be adding a full set of genes from the donor so the clone will match him or her exactly. After the cell donor's DNA genes are added to the now blank ovum , the egg will now be a fertilized egg and will start going through cell division and growth and form the fetus, and eventually become a living human baby. This will be going on inside the uterus of the surrogate mother hired to carry the child and then the baby will be born.

The sex of the clone will be the same as the cell donor. That only makes sense because we are making the clones for spare parts for the cell donor if they ever need something. Whatever the donor might someday need we can remove from the clone. And the organ to be transplanted will match the donor perfectly and never be rejected. The clone may die if the organ removed from it is vital to its staying alive, but the organ recipient may keep on living for a long time.

Would that be murdering the clone? Not really because I will create the clone so that its body is healthy and alive, but there will be no cranial brain. A person does not need the brain to stay alive as long as the brain stem is intact and that is where my expertise is important.

I know how to disable the DNA of the embryo that normally would cause the cranial brain to develop. It will be not be there, so the clone will be laying on the bed or table in a coma state. Living but not conscious. Many people live for years in such a state. In a coma, but they are brain dead. We don't need the brain to stay alive. Just the brain stem that connects above the spinal cord.

As far as I am concerned, when we kill the comatose clone to remove an organ, it is not murder because it was never conscious, but it is hard to judge the legality of the act. I will not be the surgeon who removes the organ and kills the clone. I create the embryo. I am an authority on the DNA and after I create the embryo, I will no longer be involved.

3. Growing the clone and harvesting organs

How I make an embryo for a clone is not important for you to know. It takes a long time to even understand the procedure let alone being able to actually do it. A regular person would not be able to do this. In fact, very few people in the world are capable of making any human clone and I am one of the few capable of making a human clone with no brain that will develop.

I spent many years training to understand the human chromosomes. The key to this whole thing is that we don't want the clone to have the normal cerebral hemispheres that compose the conscious brain. My clones will not be thinking. No movements. No speaking. No vision or any senses. Just laying motionless and no consciousness. That is important because someday it will used for some spare parts. We just want a minimal amount of life. Breathing and heartbeat. The vital bodily functions will be approximately normal.

As I inferred earlier, having clones of yourself will cost a lot of money. They have to be cared for in a laboratory setting. Technicians required to be on duty. High paid technicians. Physicians will have to be involved to examine the clones to make sure they are functioning properly and if they are healthy if the event arises that an organ will have to be removed.

There are some organs that the removal will not result in the death of the clone. Also some structural items like knee cartilage and ligaments for example. The clone will not experience any pain or be aware of any disability. It will continue to lay in a comatose state. If a major organ necessary for survival is removed and used, then the clone will die and be disposed of. However, many other crucial organs can be removed and preserved for future use if necessary.

That is the end of the story of that clone. Not really very exciting. Just a function of growing spare parts and harvesting them when needed. The clone will have to be attended to for many years. First it has to develop and then be kept alive and viable. Maybe 50 years or more. It can be done if the donor has a lot of money to spend. Some people do have a lot of money. Most of us don't, so most of us will never have a clone growing new organs and structures for us. We will live normal lives and if we need an organ transplant, we will wait in line for availability and hopefully it will be a genetic match that can adjust to our bodies. And we have to adjust to the new transplant also.

The physician who replaces the organ in the cell donor's body will be in a gray area of possible illegality so he or she will probably expect to be paid at least a million dollars because of the legal issues. It is possible that a surgeon might be found who would do it. Maybe someone who has been suspended from their medical practice. Money talks. It can be done for the right price. Replacing a vital organ is a somewhat risky procedure. Not a sure thing. The cell donor is going to have his or her organ removed and replaced. He or she might die during the procedure. Hopefully, it will eventually be made legal so hospital type setting can be used for the surgery. Sort of like when abortions were made legal, many lives of women were saved.

Some transplants are almost regular procedures, like kidney transplants. Doing a liver or heart, or lungs, or pancreas can be done, but way more difficult. Surgery like that is a very high technical skill that takes a long time to learn, but at least the organ being transplanted will be almost like new and if the procedure is successful, then the recipient will have a new lease on life.

Obviously, the cell donor who provided for the clone would hope that a transplant will never have to be done. Having the clone is like a good insurance policy. Hopefully it will never be needed, but if it is needed, it is available. A perfect match in every way. A perfect fit and a perfect match with the immune system so there is no chance of rejection. Just in case it is needed. Hopefully, it won't be needed.

4. An unusual request

As a scientist, I do not make judgements as to a client's purposes. If they pay me what I ask for then I will do my work and prepare the embryo to be used to grow into a clone. I don't get involved in what happens to the clone. At least not usually. Occasionally I get a request that makes me consider if what I am doing is ethically acceptable. Then I might or might not get involved in the project.

A guy contacted me that he was in love with his daughter, but he would not take a chance of ruining his family life by trying to have sex with her, although he was very attracted to her that way. I am not saying he is normal. To me, incest seems abnormal and it doesn't matter if she is underage or not. It is still incest. Of course, if she is underage then the offense is also pedophilia. It happens a lot. Sometimes with more than one girl in the family. I did know of one case that the mother had sex with her underage son. I'm sure it happens a lot. Not as much as with father and daughter.

I can make a clone that is intact physically but with no consciousness. If he brought me a scraping from his daughter, I could duplicate her and the clone would have no consciousness, no sexual desire, no rejection of sexual attempts. I don't know if she would experience arousal like vaginal heat and moisture. If those functions are controlled by the brain as is usually assumed, then she would not have any sexual reactions. But we don't know if there are some direct nervous connections that don't have to go to the brain.

We just don't know because it has never been done as far as I know.

So would that give him the kick he was seeking? I don't know the legality of it because like I said, it has never been done. The clone would look identical to his daughter and he could do his fantasy love making without disturbing the daughter and his family. His incest would be all in his mind. It would almost be like masturbating while dwelling on visions of his daughter. That is sort of like incest also, but I guess not really.

Of course, I can also make a clone with the brain intact. Everything will work. Just like on his real daughter. What about love? Does that work with a clone? We don't really understand how love works. The nerve pathways start to develop in infancy being in love with the mother. As we get older those love nervous pathways continue to develop. Even love with the father, siblings, a pet, nature, God, friends, studies, a special person, and her own kids and home. Boys the same way.

I never had a sister or a daughter so I cannot understand incest from personal experience or desire. Pedophilia was never my thing either. I like full-grown women. The way they are shaped. They way they move. And usually I like the way they talk. That is my personal turn--on, but as a scientist I understand that some people are subject to pedophilia desires. I don't like youngsters to be abused in any way. We don't have much experience with clones. As far as I know, there has never been a human cloned. But I am sure it could be done.

The lawmakers keep it illegal because they really don't understand it and they fear that clone making would get out of control. We would clone Michael Jordan and every team would have one. Same with all sports and musicians and movie stars. Artists. Everything perfect. Even a perfect prostitute. Perfect scientists. No doubt I would get replaced by a scientist cloned from a master scientist. I'm not sure how I feel about all that. My specialty is comatose clones for spare parts, but I'm not sure how I feel about clones with brains and personalities.

5. Cloning for incest or pedophilia

It can be done, but that does not make it right. I could make an exact copy of that guy's daughter. It would not be his daughter exactly because there is more to a human than just looks and basic personality. The clone would look and sounds like his daughter and probably move like her and probably even laugh like her. I have not decided how I feel about a clone being used for incest or pedophilia or really for any sexual perversion. I suppose most any abuse could be considered sexual. It is not something that we fully understand. Sexuality. There are lots of theories and many people seem to have learned opinions, but that does not mean that they really know about it.

Because some rich guy paid me to create a clone and he is paying to have it raised, does that mean that he owns the clone if the clone has a brain and a mind of its own. Does he own it body and soul? And the clone will not have any free will. In this case the clone will have a brain and a mind and a personality. What about a soul? Would unscrupulous people use the cloning to create slaves? What about sexual slaves? Would clones have any rights? Under the law. Ethically. These are all questions that must be answered before it will be legal to create and raise a thinking clone.

Would the guy who wanted a clone for incest be satisfied by sex with the clone. How much of his incestuous feeling could be transferred to a clone of his daughter and how much of the incestuous desire would only be satisfied by his actual daughter with all her subtleties of who she is. Pedophilia is a different thing than straight incest. They might overlap and they might not. Sometimes incest is its own thing. I can't explain it. To really understand it you have to have experienced it.

I think cloning for spare parts with a clone with no real brain should be legalized. Of course, there are so many people that there would be no way possible to have clones for everyone being cared for so it would just be a perk for the wealthy and that is not something I really believe in. I believe in equality of all people. Equal rights and equal opportunities. That is what I believe in, but I know it is impossible to obtain. A set of clones is a good example. The cost of making and maintaining clones is way more than most people could afford. It is only for the wealthy.

Would I want a clone for myself? Sure. Especially one with no brain. Just in case something goes wrong with an organ malfunctioning. A person might die from one organ malfunctioning, but if it could be replaced, it would add a lot of years to their life. Nothing wrong with that. I wish it could be for everyone.

At least maybe we could develop a universal spare parts clone that maybe is not a perfect fit for us like parts from our personal clone, but maybe we could develop clones that could work most of the time. Maybe large laboratories with thousands of clones waiting for an organ harvest. Maybe the clone embryos could be developed with organs that would not cause a massive immune response. Kind of neutral. That could be used by many people. I would like that. Those organ donor clones would of course, not have a brain.

What if a person needs something replaced in their brain? Then the clone would necessarily have a brain, but somehow no personality or individuality, because it may be sacrificed for its brain parts to be harvested. It could be done. Just not yet with out current state of knowledge. For one thing we dot know how to create a clone with a healthy brain that is not activated. It would not be right to kill a fully functioning clone to take some of its brain.

6. Right or wrong?

How do I feel about cloning being illegal? Well, like you can infer, I consider cloning with no brain to be a good thing and should be made legal. But of course, it contributes to elitism which I consider to be an evil expression of humanity. So, I have mixed feelings about the comatose cloning at the present time. It would be good for learning. A good experiment. But eventually there will have to be a more generalized clone that can contribute organs to anybody that needs them and not to just the wealthy.

Cloning with the brain DNA left intact is a whole different issue. The clone is more completely alive. A real person. It can be raised in a home and doesn't necessarily have to be confined to a table or bed in a laboratory setting. I think it would be a good thing if the clone was going to be adopted and the embryo might even be implanted in the uterus of the woman, whether she is single, married to a man, or married to a woman. Just so the clone is going to have a home. Wouldn't be for spare parts or sexual gratification. It would be a child for a person that wants it.

If the embryo was made from the mother's egg and the father's sperm, then it would not be a clone as such. It would basically be artificial insemination where the embryo is created in the laboratory instead of in her womb. But the result would be the same. An embryo with both the mother's and the father's genes. You might say it is normal reproduction with a little help from science.

A true clone would be made by taking the genetic material out of the egg which could be obtained from the future mother or from an egg donor. Often for one reason or another, the mother's eggs are defective possibly from age, and it would be better to use an egg from a young egg donor and remove that donor's genetic material so you have a blank ovum ready to be fertilized with DNA either from the father and mother, the father alone or the mother alone, or two women, or two men. All those combinations are possible. They haven't been done yet, but they will in time. Those are all variants of cloning.

My particular interest is with the spare parts comatose clone. Not so many ethical questions. Legal questions yes but the main ethical question concerns giving wealthy people even more of an advantage in life. Keep raising clones, harvesting organs and living an indefinite length of time. Only for the wealthy unfortunately. I'm not sure I can be part of a system like that. But I do like the idea behind extending the human lifespan with replaced organs grown from clones.

What about multiple identical clones? There would have to be a good reason that I can not think of right now. I can see a single clone for people who a re going to raise it and always care for it like their own child. Which basically it really is. In that situation, I can see cloning be morally right and legal. If lawmakers lifted the restrictions on cloning. But I think that many lawmakers are considering that clone makers would make multiple copies of a clone. They don't like that and I don't think that I do either.

7. Cloning multiples

Obviously different kids from the same father and mother are not always the same. Even identical twins have some differences. The genes on the chromosomes mix differently so you get siblings who can be very different. In cloning if we take cells from the same person to make the clone, it does not mean the clones will be identical for sure, but probably they will be very close to identical. If you wait until the cloned baby is somewhat developed and then draw some cells from its skin, then that would be good way to get identical clones if that is what you are after. If you just use cells from an adult person, there is a chance that the chromosomes will be slightly mutated and that would result in clones that are slightly different.

Why would you want clones that are identical? If they are being grown for spare parts, it is obvious you want all the clones to be identical to the original cell donor, but what about if the clone is grown with an intact brain and is walking around. Talking and thinking. What could possibly be an advantage to develop clones like that. They wouldn't be used for spare parts because they are alive and have a right to keep their own organs. Killing one of them and taking the organ would be akin to murder.

So making identical clones with intact brains is sort of disdained by scientists even though they can probably be done fairly easily. You would not be just creating spare parts. You would be making a group of identical people. I guess it would be sort of like identical twins or triplets or quadruplets, but even closer to being the same. The question is, why would it be done?

Well, obviously it could be an ego trip for some wealthy person to want more of themselves populating the Earth. One of them is not enough. They think they are very special and maybe they think their physical and mental character should be duplicated for the value to humanity. Yes, some people do think that way. It is possible, but probably it is not the reason multiple clones would be created. Probably there would be a financial benefit for the procedure. To make a copy of a person with some unusual and valuable characteristic. An ability that can be duplicated profitably.

One factor that would have to be considered before identical clones are created is these are going to be identical intelligent individuals with functioning minds. They cannot necessarily be controlled. There can be no ownership of clones with their own brains intact. That would be too much like slavery which should be strictly forbidden.

The problem is humanity cannot be trusted. There is still way too much evil in the world and there is too much of a chance that the clones would be abused. They might not be given opportunities to make choices for anything in their lives. Just created to do a function and not to have an actually enjoyable life. These are complete humans even though they are clones. They are entitled to the same basic rights as any other person and that means in America or other Western countries. Not in a country where they might be held in bondage against their will. Or possibly with their will being withdrawn.

How can the technology be controlled? To make sure cloning does not get into the hands of people who will abuse them for their own profit. Say ten clones doing some miserable job with just enough remuneration to barely survive. There are people like that in the world who would do that. A lot of them. They cannot be allowed to have clones. Maybe the world is not ready for cloning. Clones with brains are real people. They deserve civil rights like everybody else.

And the same goes for their sexual rights. They can't be told when to have sex. They can be asked, but they have a free will choice. Just like any other person. So that rules out the pedophilia people and the incest people. At least in my opinion.

It will be hard to keep the pedophiles and incest cravers out of cloning completely because they are numerous and sometimes very wealthy. It is likely they can find an unscrupulous scientific technician who will make them a clone without any safeguards that the clone will have a somewhat normal life. It won't be me. I will only make clones without brains or if the clone is to have brains, I will need to be assured it will have a good home.

8. Clone fantasies

I can't tell any stories about real life experiences with human clones because there have never been any yet. At least none that I know of. Any clone stories would have to come from my mind and just be a fantasy. Naturally since I have studied so much about clones and done so much work developing clone creation techniques, that I do have fantasies about what life with clones would be like.

I have thought about what it would be like to be married to a clone whether I was involved in her creation or not. Being a bachelor, I do think about getting married and why not marry a clone. She naturally would be a lot younger than me because I would have to wait until she was at least legal age which is probably 18.

I would not try any pedophile stunt on her. I want her to be a matured woman and ready for marriage. She can still finish her education while she is married. I don't think I want to wait until she is done going to college to marry her because I am getting older myself. I would wait until she is 18 although I hope I can know her non-sexually as she is growing up.

Maybe she could be raised by a woman friend of mine who I approve of. She would not have been created from my DNA so our love would not be incestuous even though I might have some influence in guiding her development. Not forcing my will on her. Just making subtle suggestions that may or may not be followed. I might fall in love with her as she is growing up. But I would never express it sexually.

If she was a multiple clone, then I would get to know all of them individually as they were growing up. Again, it would not be pedophilic sexual. When the time was right, and she was of legal age then I would try to court her and propose to her to be my wife. If she was a multiple clone, the choice would be more difficult getting to know which clone that I loved and wanted to marry. They would be so much alike. What a fantasy it would be for me. Hopefully there would be true love develop between one of them and me.

I'm not even going to think about what it would be like to have affairs with two or three of them before we got married. A taste of honey with each of them. How could I not fantasize about that possibility? Would there be a difference in the way they made love. Would I even know which one I was with? It would be exquisite pleasure especially if they all accepted sharing my attentions.

It is not good for me to even think about that happening. What an amazing temptation that would be? I can't allow myself to be drawn into that fantasy because I am scientist and I could make it happen. I would just need to find a woman to help raise them. Maybe I would make the clones from her DNA so she would be more inclined to care for them.

This fantasy can go on and on and get out of control and make me a slave to my fantasies. Actually, any one of the women would be adequate for my needs. It is just a guy's type fantasy that multiple lovers would provide an endless variety of sexuality that would never wear out. I know it is not true. One is plenty for me. One perfect love.

9. Robotic clones

I have already stated that brain intact clones are independent people and cannot be just used to satisfy someone's whim. The comatose clones don't really have much responsiveness. Probably none. They could be used by a pervert with no ethical violation and probably no legal violation. But there is the issue about responsiveness. Can the perpetrator get the satisfaction they require if the clone does not move or have any type of response? It would be sort of like the sex dolls that are sold in adult shops.

A male comatose clone would not get erect but could be used manually or orally if that is a kick. I don't think it would. Not the male or the female comatose clones of any age would have a lasting sexual attraction for the pervert. They want some kind of response. To know it is a real person, even if underage. There could be some people with sort of a necrophiliac connection that the clone's lack of responsiveness would be an added attractiveness, but not for most of the pedophiliacs.

So the comatose clones are not usually good sexual partners. Certainly not for on a regular basis. The brain intact clones are off limits morally and legally because they are regular people with rights and their own desires and needs. Is there a middle ground? What about a robotic clone? It would be moving around. Seeing. Speaking. Sexually responsive. I could make the embryo without that part of the brain associated with determining their self-determination. I guess you could call it their ego. That part of their mind would not exist.

But they would have personality, charm, humor, rudimentary emotions, feelings that flow like anyone else, and be subject to sexual arousal. Would be very similar to a complete person but no thoughts of future or plans or involvement in memories, and not affected by any memories. No sense of right or wrong, no expectations, no demands, no feelings of love. Love would require way more complex emotional contemplations than the robotic clone is capable of.

Probably all sexual activities would be acceptable to a robotic clone. Really about anything goes. They would have all the senses which means pain might be a factor unless the clone could be manipulated and trained into liking pain and maybe even desiring it. There are not many limits to what a robotic clone would and would not accept or could learn to accept.

I personally can see how it could be considered immoral to have sex with a robotic clone. It would be your sex slave because it cannot live independent from you or someone. It needs a master. It needs to be controlled. The robotic clone does not have enough of its brain functioning to think for itself. It reacts and lives in the moment. No thoughts of consequences. The robotic clone is almost on a perpetual high. Just go with the flow.

As usual, legality if another story. Lawmakers always want to have a say in what you can and cannot do. Probably that is a good thing most of the time because humanity is certainly not universally ethical. So laws would have to protect the clones from abuse. For many generations to come, the clones will not even be legal, but eventually controls will be able to be legislated and regulated. It will sort of be like treatment of pets, but even more tightly controlled. They will be pets of a sort. Human pets. Not slaves. Pets.

10. Clone reproduction

As time goes on clones will be legalized and produced commonly. All the types of clones will be fertile unless purposely sterilized. That is possible. It could be part of the procedure in making the embryo to disable the clone's ability to make sperm or ova. That can be controlled in the DNA. Right now I don't know how because I haven't thought about it and it would require considerable study and experimentation. But I'm sure it could be done. Intact brain clones might be able to reproduce and raise their kids.

Someday clones and their children and grandchildren will interact in the society in general and people probably won't know if a person is a clone or descended from a clone. It would be hard to know unless their chromosomes were examined by DNA testing. Every cell has two sets of chromosomes. Normally one from the mother and one from the father, but with a clone all the chromosomes are from the cell donor and the two sets would be from his or her mother and his or her father.

It doesn't make much difference. The clone will still produce sperm and ova with the proper number of chromosomes which were derived from the cell donor's chromosomes which were derived from his or her mother and father. It just sort of adds an extra generation in the process because in normal reproduction the sperm and ova are created right from the person's own chromosomes. With cloning, the same sperm and ova will be created by the clone. Same genetic material as the original donor's sperm and ova. Just a generation later.

A person could marry a clone and have children with that clone and maybe not even know their spouse is a clone and unless the clone chose to tell them. Someday people might say something like "my grandmother was a clone" sort of like they say now when they are saying they had an ancestor who was Native American. The clones will be integrated into the society and no one will be able to tell if someone was a clone or descended from a clone. They will be like everyone else. Unrecognizable as such. Even with genetic testing. It is possible though that clone makers of the future may add a "genetic marker" to the DNA. Some segment that has no purpose except to identify the DNA as belonging to a clone and it would be passed to their descendants. Just another possibility.

A clone could marry another clone who is from a different cell donor. No one would even know, unless they chose to tell people. Clones will be such perfect replicas that no one will be able to tell if a person is a clone or not. Of course, if they were cloned from someone with exceptional good looks or special talents or skills then the clone will also have those looks and skills. One hundred years from now cloning will be common. People at the top of the society will either be clones or have clones in their lineage. Some people produced by regular reproduction may be considered kind of old-fashioned. To be a clone or to have clones in your family may become a very honorable thing.

www.ingramcontent.com/pod-product-compliance
Lightning Source LLC
Chambersburg PA
CBHW070902220526
45466CB00005B/2095